NASA's ARTEMIS MOON MISSION: Facts About Artemis I Launch

ROGER MUNIZ

Table of Contents

Introduction

When the uncrewed Artemis I mission launches on Monday, August 29, it's simply the first step toward the future of space exploration. The last crewed landing on the moon, Apollo 17, occurred over 50 years ago. The last Apollo mission's record for the longest crewed deep space voyage still stands at 12.5 days.

Through the Artemis program, which seeks to put people near the undiscovered lunar south pole and ultimately on Mars, astronauts will embark on long-duration deep space missions that push all of the limitations of exploration.

Why NASA is returning to the moon 50 years later with Artemis I

"We're going back to the moon to learn to live, to work, to survive," stated NASA administrator Bill Nelson during a press conference earlier this month.

"How can you keep people alive in such harsh conditions? And we're going to learn how to utilize the resources on the moon to be able to create things in the future when we move — not a quarter of a million miles away, not a three-day voyage — but millions and millions of miles away on months and months if not a years-long journey."

NASA astronaut Randy Bresnik emphasized the need of utilizing lunar exploration as a means to prepare for landing on Mars at a NASA briefing on Saturday.

When camping in the Alaskan wilderness, you wouldn't only depend on fresh gear and shoes that haven't been broken in yet, he added. Mars isn't the place to try out new gear for the first time, however.

"We're going to go to some local spots a bit closer initially," Bresnik remarked. "Then you may return home if your shoelaces break or anything like that."

Astronauts have lived and worked aboard the International Space Station, which orbits roughly 254 miles above the globe in low-Earth orbit, for more than 20 years. Their experiences, which may last from six months to over a year, have shown how the microgravity environment affects the human body.

"Every day that I spent on the space station, I looked at it as walking on Mars," said NASA astronaut Reid Wiseman, Chief of the Astronaut Office at Johnson Space Center in Houston. "That is why we're up there. We're trying to make life better on Earth and we're aiming to spread mankind throughout our solar system."

On Artemis II, slated for 2024, astronauts will travel the same course as Artemis I — circling the moon at a larger distance than any of the Apollo missions. Artemis III, set for late 2025, would land the first woman and the next man on the moon's south pole, where permanently shadowed areas may hold ice and other resources that might support people during protracted moonwalks.

"Our moon serves as essentially a cosmic library right next door," said Jacob Bleacher, NASA's top exploration scientist. "Lunar rocks and lunar ice fundamentally function as the books of this library. We may utilize them to begin to illustrate how the solar system has developed. This can assist us to get insight into what was occurring here on Earth when life was gaining a footing in the solar system."

The Artemis mission envisions establishing a continuous human presence on the moon and placing an orbiting lunar base dubbed the Gateway in situ.

This picture illustrates SpaceX's Starship human lander concept that will deliver the first NASA people to the surface of the moon under the Artemis mission.

"We want to remain on the lunar surface and learn on the lunar surface so that we can gather the greatest science and know how we're going to get to Mars," said Jim Free, assistant administrator for NASA's Exploration Systems Development Mission Directorate. "On Apollo, we accomplished tremendous science near the equator. This time, we're headed to the South Pole."

Over time, the SLS rocket will develop, Nelson added. By the time the Artemis IV mission rolls to the launchpad later this decade to dock with the Gateway, the rocket will be taller and even more powerful than the version used for Artemis I.

Artemis I will bring the first biological experiment to outer space

Artemis I is a test mission, Nelson emphasized. It serves as the initial flight of the Space Launch System Rocket, the Orion spacecraft, and its heat shield, as well as protective gear for future astronauts and evaluating radiation exposure.

A series of scientific experiments and technological demonstrations within Orion and flying outside of it on tiny satellites called CubeSats will collect new data on the space environment future Artemis passengers will experience.

The lessons learned from Artemis I, which will be gathered when it splashes down in

October, might shape the following phases of the Artemis program.

Currently, the first five Artemis missions have been planned, and NASA is working on setting out the specifics for missions six through ten, Free added.

Teams at NASA are "working through the broad exploration goals and then refining down to an architecture which gets us out to Mars," Free added. "We're looking at rolling through that architecture, choices, and process in the early part of next year."

The objective of landing people on Mars by 2033 was established by the Obama Administration, and NASA executives have embraced the aim ever since.

"With the Artemis, I launch on Monday, NASA is at a historic inflection point, set to commence the most important sequence of scientific and human exploration missions over a generation," said Bhavya Lal, NASA assistant administrator for technology, policy, and strategy.

Why NASA is returning to the moon 50 years later with Artemis I

Almost 50 years after the last Apollo mission ventured to the lunar surface, NASA has established a program that promises to land humans on unexplored lunar regions and eventually on the surface of Mars — and it all begins with Artemis I.

It's no coincidence that the Artemis program is named for the twin sister of Apollo from Greek mythology. Artemis will pick up where the famed Apollo program left off in 1972 by sending crewed missions to the moon, but in a new way.

The goals of the Artemis program include landing diverse crews of astronauts on the moon and exploring the shadowy lunar south pole for the first time. The ambitious program also intends to establish a persistent presence on the moon and construct reusable technology that might allow human exploration of Mars and potentially beyond.

But none of this is feasible without first taking one giant leap. When Artemis I launches, the uncrewed mission will test every new component that will make future deep space travel viable before people undertake the trek in 2024 and 2025 onboard Artemis II and Artemis III, respectively.

Chapter 1

What Will the Nasa Moon Mission be Transporting Into Space?

At three meters tall, Nasa's Orion spacecraft is roomier than Apollo's capsule and seats four astronauts instead of three, but during Monday's test flight it will contain cargo ranging from a mannequin called Helga to components of Apollo 11's engine and the occasional stuffed toy.

For the flight, a full-sized dummy in an orange flying suit will occupy the commander's seat, outfitted with vibration and acceleration sensors. The "commander" was dubbed Moonikin Campos in a public contest, in honor of Arturo Campos, a Nasa engineer who helped rescue Apollo 13 from tragedy by finding out how to jury-rig its half-broken electrical system to return the crew home.

Artemis 1: Two additional mannequins constructed of material imitating human flesh — heads and female torsos but no limbs – will detect cosmic radiation, one of the main threats of spaceflight. They are called Helga and Zohar. One torso is trying on a protective garment from Israel.

Unlike the SLS rocket below it, Orion has launched previously, performing two loops around Earth in 2014. This time, the European Space Agency's service module will be connected for propulsion and solar power through four wings.

Besides three test dummies, the voyage contains a bevy of deep-space research projects. Ten shoebox-size satellites will blast off once Orion is rushing towards the moon.

These "CubeSats" were inserted in the rocket a year ago, and the batteries for half of them couldn't be recharged as the launch kept being postponed. Nasa anticipates some to fail, given their low-cost, high-risk character. The radiation-measuring CubeSats should be OK, coupled with a solar sail demonstration targeting an asteroid.

Orion will carry a few slivers of moon pebbles gathered by Apollo 11's Neil Armstrong and Buzz Aldrin in 1969, and a bolt from one of their rocket engines, rescued from the sea a decade ago.

Aldrin isn't attending the launch, according to Nasa, but three of his old colleagues will be there: Apollo 7's Walter Cunningham, Apollo 10's Tom Stafford, and Apollo 17's Harrison Schmitt, the next-to-last man to walk on the moon.

Snoopy plush toys are a Nasa tradition, coming from the name of the Apollo 10 lunar module that traveled to the moon to test descent and landing procedures but never got to land itself.

A Snoopy toy rode aboard the Columbia space shuttle, and this time the cartoon figure will be joined by Shaun the Sheep, to celebrate the cooperation of the European Space Agency. Their official goal is to show zero gravity, by floating about.

The Observer opinion on the Artemis deep space project: $93bn? Worth every dime Observer editorial Orion will carry Biological Experiment-01 featuring tests on seeds, fungus, yeast, and algae. Also on board is a speech recognition demonstration dubbed Callisto, built jointly by Amazon, Cisco Systems, and Lockheed, that will demonstrate how Amazon's Alexa functions in a space capsule and how such a system may be used by future astronauts.

The fake astronauts' official "flight kit" contains hundreds of additional things, many of which will become "flown in space" memories back on Earth. They contain seeds that will be sown to become "moon trees", a Dead Sea stone, mission patches, stickers, USB drives, and country flags. Some Lego has even made it aboard.

Chapter 2

Facts About Artemis I

The Space Launch System rocket that will transport an uncrewed Orion spacecraft to the moon in the Artemis 1 mission is the most powerful rocket NASA has ever developed. It weighs 5.75 million pounds when fully fuelled for takeoff, but will ascend almost 500 feet straight up in only seven seconds.

Here are some astonishing facts and data concerning NASA's powerful rocket:

Fuel Load & Shrinkage

When filled with 537,000 gallons of liquid hydrogen at minus 423 degrees Fahrenheit, the massive fuel tank in the SLS core stage will shrink by around 6 inches in length and 1 inch in diameter. The rocket's liquid oxygen tank will reduce by an inch and a half in length and around 1.3 inches in diameter. Because of the shrinkage, everything that attaches to the tanks — ducts and vent lines, brackets, etc. — must attach using accordion-like bellows for flexibility.

Retro Processor

The core stage flight computer, which controls all elements of the rocket's rise to space, employs the same kind of Power PC CPU as a long-out-of-date G3 Macintosh Powerbook. The customized operating system, however, is significantly more efficient.

Jumbo Jets

The 2 million pounds of power from the core stage's four RS-25 engines could maintain eight 747 jumbo airplanes in flight. The 3.6 million pounds of thrust from every two solid-fuel rockets could propel 14 four-engine jumbo planes.

Powering Up

The energy output of the RS-25 engines, if turned into electricity, would power roughly 850,000 kilometers of lighting on a road spanning to the moon and back and then 15 times around the Earth. The four engines offer double the power required to move 10 Nimitz-class aircraft carriers at 30 knots.

Guzzling propellant, all four core stage engines burn 1,500 gallons of propellant every second, enough to empty an average 20,000-gallon swimming pool in 13 seconds.

The SLS rocket's two solid-fuel boosters use 5.5 tons of propellant every second. The heat created by the boosters during their two minutes of operation, if turned into energy, would power 92,000 houses for a whole day.

Fierce Heat

The rocket booster exhaust is hot enough to fuse desert sand into the glass during test firings at Northrop Grumman's factory in Utah.

Bolt Blasting

The 5.75-million-pound SLS rocket is secured to its launch pad by eight huge bolts, four at the base of each solid rocket booster. At booster activation, explosive charges fracture the fasteners and the SLS lifts flight.

Acceleration

In the first two minutes of flight, the twin solid rocket boosters produce 75% of the SLS rocket's total power, blasting the spacecraft to over 4,000 mph and an altitude of 27 miles.

When the rocket's main engines shut down eight minutes after liftoff, the rocket will be traveling at about 18,000 mph. That's quick enough to traverse 88 football fields, set end to end, in one second. The rocket's upper stage will raise the velocity to 22,600 mph — 110 football fields per second — to break out of Earth orbit and launch the Orion capsule to the moon.

Chapter 3

What Does NASA Aim to Learn from an Uncrewed Journey to the Moon?

The huge rocket is vital to NASA's Artemis program. This mission, designated Artemis 1, is the first of three planned Artemis missions that will conclude in 2025 with astronauts stepping foot on the moon for the first time in 50 years and will feature the first woman and person of color ever to do so.

Ultimately, NASA intends to build a permanent lunar base at the moon's south pole, serving not only as a residence for moon-bound astronauts but also as a staging ground for crewed missions to Mars and deep space exploration, Pat Troutman, Strategy and Architectures Liaison for NASA's Moon to Mars Architecture Development office, told Live Science.

However, every journey of a thousand light-years begins with a single step — and Monday's premiere launch of the Space Launch System (SLS) rocket (also known as the Mega Moon Rocket) will be unscrewed, with only three mannequins riding aboard the Orion Crew Capsule perched atop the rocket's tip.

"This is the first flight of a significant space system," Troutman added. "It's a highly connected, sophisticated system with plenty of energy, and normally you want to test things the first time without anyone too near."

According to Troutman, the Artemis I mission will largely test two things: The performance of the SLS rocket and Orion Crew Capsule, and the safety of the humans within.

For the sake of this expedition, the astronauts will be portrayed as three mannequins — or "manikins" — traveling inside the Orion spacecraft.

Sitting up front, Commander Moonikin Campos (named after former NASA scientist Arturo Campos, a significant player in the Apollo 13 mission of 1970) will try out NASA's new space suit, the Orion Crew Survival System flying suit. Behind him will sit Helga and Zohar — two "phantoms," or limbless mannequins comprised of "materials that replicate human bones, soft tissues, and organs of an adult female," according to NASA. (Commander Campos' name was decided in a public contest; Helga and Zohar were named by the German and Israeli space agencies, who are partners on the project).

Campos and Zohar will wear special vests to shield them from the powerful solar radiation that Earth's atmosphere generally absorbs; the third mannequin will go vestless to act as an experimental control (sorry, Helga).

All three mannequins will sit on chairs fitted with sensors to detect the acceleration and vibrations during the spacecraft's launch and descent to Earth. By evaluating the manikins and their sensor data after the mission is complete, NASA should have a clear picture of the possible physiological strain and radiation exposure that human astronauts might anticipate undergoing during future stages of the Artemis program.

The Space Launch System (SLS) is the Most Powerful Rocket Ever Constructed

The third key event is the trans-lunar injection – a vital maneuver that lasts around 20 minutes, whereupon the now-booster-free spacecraft uses a smaller

RL10 engine to push fully out of Earth's orbit and start on a track towards the moon. Five days later, the Orion spacecraft will be on the moon's doorstep, circling within approximately 62 miles (100 km) of the lunar surface.

After many weeks of circling the moon, collecting photographs, and completing tests on different spacecraft systems, the Orion capsule will return to Earth. This puts in action the last high-energy event: the violent fall through Earth's atmosphere, during which the spacecraft will suffer temperatures of roughly 5,000 degrees Fahrenheit (2,760 degrees Celsius) — nearly half as hot as the surface of the sun.

Finally, the capsule will release parachutes and splash down into the Pacific Ocean off Baja California, Mexico. How the spacecraft endures through these high-energy occurrences will inform NASA if the Artemis program is ready to move to its second phase. In Artemis II, presently scheduled for May 2024, a team of actual human astronauts will repeat the voyage around the moon that their mannequin colleagues went on during Artemis I.

The safety and success of Artemis II rest on what scientists can learn from Monday's launch, and from the 40 days that follow.

Chapter 4

What Advantages will Artemis Provide?

With Artemis missions, NASA will put the first woman and first person of color on the Moon, employing revolutionary technology to explore more of the lunar surface than ever before. We will engage with commercial and international partners and develop the first long-term presence on the Moon. Then, we will utilize what we learn on and around the Moon to take the next major leap: sending the first humans to Mars.

Why Are We Going to the Moon

We're going back to the Moon for scientific discoveries, economic rewards, and inspiration for a new generation of explorers: the Artemis Generation. While retaining American leadership in exploration, we will form a global alliance and explore deep space for the benefit of everyone.

Astronauts on Moon

With Artemis, we're drawing on more than 50 years of adventure expertise to revive America's enthusiasm for discovery.

Blueprint of Space Launch System

Artemis missions promote a burgeoning lunar economy by igniting new sectors, supporting employment development, and expanding the need for a trained workforce.

Blueprint of Mobile Launch Crawler

We will explore more of the Moon than ever before with our commercial and international partners. Along the way, we will connect and inspire new audiences — we are the Artemis Generation.

Our Success Will Change the World

We will create an Artemis Base Camp on the land and the Gateway in lunar orbit. These characteristics will enable our robots and astronauts to explore more and perform more research than ever before.